Exotic Animal Care and Management Student Workbook

Vicki Judah AS, CVT
Kathy Nuttall, BS, CVT

Australia • Brazil • Japan • Korea • Mexico • Singapore • Spain • United Kingdom • United States

Exotic Animal Care and Management Student Workbook
Vicki Judah, Kathy Nuttall

Library of Congress Control Number: 2007044864

ISBN-13: 978-1-4180-4200-4

ISBN-10: 1-4180-4200-5

Delmar
Executive Woods
5 Maxwell Drive
Clifton Park, NY 12065
USA

Cengage Learning is a leading provider of customized learning solutions with office locations around the globe, including Singapore, the United Kingdom, Australia, Mexico, Brazil, and Japan. Locate your local office at **www.cengage.com/global**

Cengage Learning products are represented in Canada by Nelson Education, Ltd.

To learn more about Delmar, visit **www.cengage.com/delmar**

Purchase any of our products at your local bookstore or at our preferred online store **www.cengagebrain.com**

Printed in the United States of America
2 3 4 5 6 7 14 13 12 11 10

Contents

Preface

This student workbook is designed to accompany Exotic Animal Care and Management. Recognizing that there are many learning styles, we have included a variety of approaches to help students become familiar with terminology and the species covered in the text. We have included case studies and ask for reflective thought. Students will benefit greatly by accessing web sites through key search words and by using the text and glossary to assist them with each section of the manual. Many references in the core text are professional papers presented at conferences. These may be accessed for further reading in specific areas of interest and to gain further insight and appreciation of the keeping of exotic species.

UNIT I

Chapter 1

Introduction to Exotic Animals

1. Research the different categories of animals listed in the CITES Appendices and determine
   a) How many species listed in each category are available in the pet trade?

   ____

   b) What impact does the trade in exotic animals have on wild populations?

   ____

   ____

Key Search Words

CITES, Endangered Species

2. Investigate the instances of unwanted exotics being released into the wild.
   a) What are some of the common species that have been identified as being released? (List 3)

   ____

   ____

   ____

   b) What problems are created when non-native species invade a habitat? (List 3)

   ____

   ____

   ____

Key Search Words

Release of non-native species, Exotic animal, Invasive species, Animal Protection Institute

3. Suppose that you wanted to acquire a pet skunk but are unsure if it is legal where you live.
   a) How could you determine if it is permitted and what special requirements would need to be met?

   Source(s) used ______________________________

   Permits: ______________________________

   Requirements: ______________________________

   b) What could be the consequences in your area of keeping an illegal species?

   Consequence for the animal ______________________________

   Consequence for the owner ______________________________

c) What are some of the reasons there are laws prohibiting the keeping of certain species?

____

____

____

____

____

d) Why are some of the laws so complicated? For example, you may find possession of a certain species is legal in your state, but not in your municipality.

____

____

____

Key Search Words

Exotic Pet Laws, APHIS, USDA (Information from the internet should not be relied on entirely as it could be outdated or inaccurate.)

Other Sources

Contact directly your community and state agencies: Division of Natural Resources; local: Department of Health, Fish, and Game.

Chapter 2

Zoonotic Disease

1. Access the website of the Center For Disease Control.

   a) Determine the incidence of zoontic diseases. How frequently are they reported?

   ____

   b) Is there a pattern to disease outbreaks? Do they occur in a specific region or locale or at specific times of the year?

   Pattern to location ______________________________

   Pattern in seasonality? ______________________________

   c) Which species of animals are most commonly implicated?

   Exotics ______________________________

   Domestic ______________________________

   Agricultural ______________________________

Key Search Words

CDC, USDA/APHIS

2. With the information gathered from the CDC and USDA, access other sites which provide information on zoonotic diseases.

   a) Determine how the information is gathered and reported.

   Define *Reportable Disease* ______________________________

   b) What is the percentage of disease transmitted by exotic species? Exclude domestic and agricultural species. To determine this, divide the number of exotic species by the total number of species listed. For example: there are 460 reported cases of zoonoses. Of these, 231were attributed to contact with an exotic animal.

   Percentage you have calculated ______________________________

   c) From this percentage, create a simple graph which reflects the routes of transmission, using direct contact, inhalation, and fomite, using one color to designate each. Overlay with other colors those that are viral, bacterial, fungal, protozoal and parasitic.

   d) What is the most outstanding feature your graph reflects? For example, does your graph reflect a high incidence of bacterial disease through inhalation transmission, a relatively low incidence of viral transmission via direct contact, or does the disease pattern seem to be level?

   e) With the results of you graph, determine the best ways to prevent the transmission of zoonotic disease. List at least eight preventative measures.

   ____________ ____________ ____________ ____________

   ____________ ____________ ____________ ____________

Key Search Words

Exotic animal disease, zoonotic potential, diseases from animals

Word Scramble

Unscramble the following terms:

Zoonotic Word Scramble

1. OTROPZAON _ _ _ _ _ _ _ _ _
2. UPALGE _ _ _ _ _ _
3. OSOESNOZ _ _ _ _ _ _ _ _
4. TNPEHAGO _ _ _ _ _ _ _ _
5. APETARSI _ _ _ _ _ _ _ _
6. AECBTIAR _ _ _ _ _ _ _ _
7. DPAMCENI _ _ _ _ _ _ _ _
8. URSVI _ _ _ _ _
9. FSNGUU _ _ _ _ _ _

4. Using your textbook, define the unscrambled terms in the space provided.

P____

P____

Z____

P____

P____

B____

P____

V____

F____

Crossword Puzzle

5. Complete the crossword puzzle of zoonotic disease related terms.

Across

3. causative agent of bubonic plague
4. single celled disease causing organism
8. Black Death
9. Food and Drug Administration
10. a single celled organism
11. virus that causes human smallpox
12. disease resembling human small pox
14. widespread or worldwide epidemic

Down

1. LCM
2. pathogen the invades host cells
5. H5N1
6. disease that can be transmitted from animal to man
7. bacterial infections that have become drug resistant
8. something that lives on or within a host with no benifit to the host
9. a dermatophyte
10. any disease causing agent
13. Center for Disease Control

UNIT II

Chapter 3

Introduction to Small Mammals

1. Make a reference list for pet owners to determine what they can feed to their small mammal herbivore.
2. Make a reference list for pet owners to determine what they can feed to their small mammal carnivore.
3. Make a list of common small mammal eutherians available in pet stores.
4. Make a list of common small mammal metatherians available in pet stores.

Word Scramble

Solve the following word scramble relating to small mammals.

1. PATSES _ _ _ _ _ _
2. ETNUREHISA _ _ _ _ _ _ _ _ _ _
3. IRVOCNREA _ _ _ _ _ _ _ _ _
4. SEENTNDLEOOK _ _ _ _ _ _ _ _ _ _ _ _
5. ORTEOMMNES _ _ _ _ _ _ _ _ _ _
6. ATERHTMIEA _ _ _ _ _ _ _ _ _ _
7. OORNIVME _ _ _ _ _ _ _ _
8. MULLSAE _ _ _ _ _ _ _
9. IRAILCALT _ _ _ _ _ _ _ _ _
10. RPEILCCOA _ _ _ _ _ _ _ _ _
11. CSIUN _ _ _ _ _ _ _ _ _
12. RRIVEHOBE _ _ _ _ _ _ _ _ _
13. ATERRTBEVES _ _ _ _ _ _ _ _ _ _ _
14. NDMLIBAE _ _ _ _ _ _ _ _
15. DODCESUIU _ _ _ _ _ _ _ _ _

Crossword Puzzle

Solve the following crossword puzzle relating to small mammals.

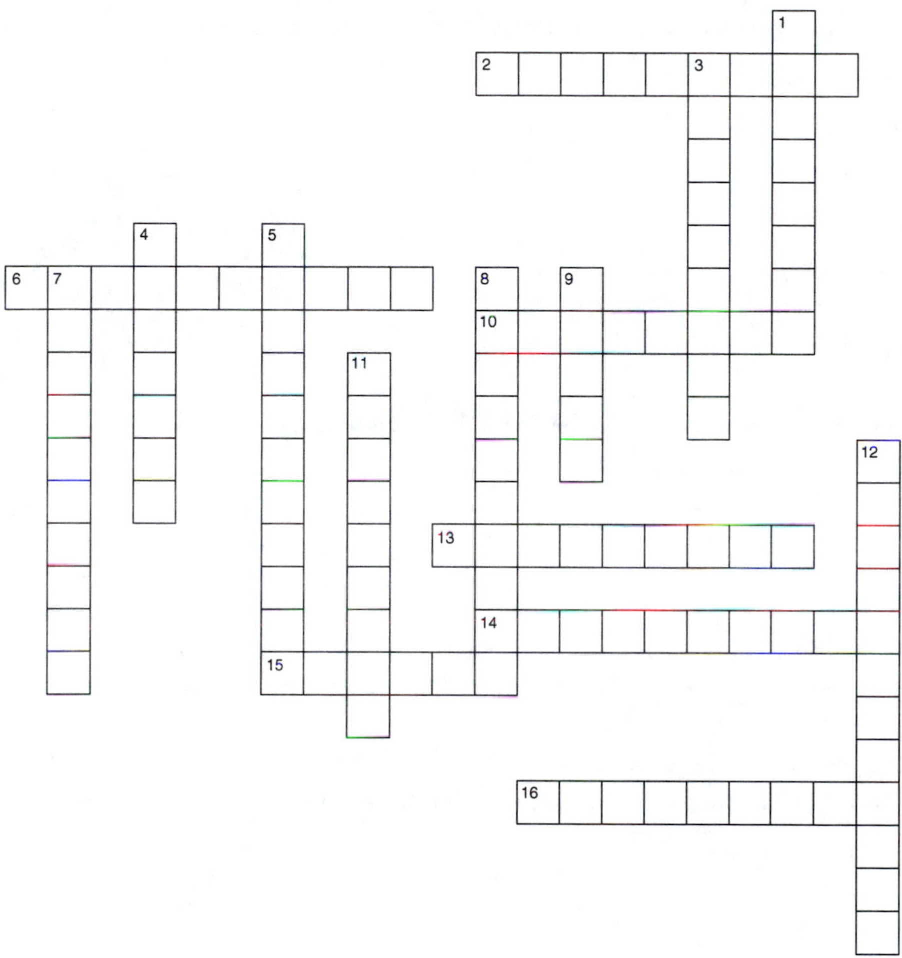

Across

2. young born blind and deaf with no hair.
6. another name for marsupials
10. mammals that eat both plant material and other animals
13. baby teeth
14. mammals with a placenta that nourishes the unborn young
15. the "stirrup" bone located in the middle ear
16. mammals that eat plant material

Down

1. lower jaw
3. animals that eat other animals
4. the "hammer" bone of the middle ear
5. mammals with a spinal column as part of their endo-skeleton
7. another name for warm-blooded mammals
8. mammals that lay eggs
9. the "anvil" bone of the middle ear
11. young that are fully developed and functional when born
12. bones that provide structure and support to the body

Chapter 4

Ferrets

1. Determine whether ferrets are legal in your area. If they are not, research the reasons they are banned.

   Legal? Yes__________ No____________

   If your answer is No, list the reasons here:

   ____

   ____

   ____

2. If ferrets are legal, visit several pet stores and answer the following questions:

   Where do retailers obtain the ferrets they offer for sale?

   ____

   You may notice two small tattooed dots in one ear of the ferret. What is the significance of these tattoos?

   ____

Case Studies

Case Study I

History: A three-year-old spayed female ferret has been vomiting for the past three days. The owner reports some weight loss and a decrease in appetite. For the last two days, there has been diarrhea in the litter box and it appears to be green.
Physical Exam: Upon examination, the veterinarian notes some muscle atrophy in the hind quarters and that the ferret is underweight. Temperature is 100.6F, Heart rate 110bpm, Respiratory rate 24bpm.

a) Given this information, what might be the problem?

   ____

b) What are some of the problems that need to be ruled out?

   ____

c) Describe how the signs and history confirm the diagnosis for the veterinarian.

   ____

Case Study 2

History: The owner of a four-year-old female spayed ferret states that the ferret has been losing hair over the past three to four weeks, but other than that, the ferret appears to be normal. The owner also reported that the pet store told her that the hair loss was "normal because they shed twice a year."
Physical Examination: There is alopecia of the tail and thinning of hair over the pelvic area. The jill has a swollen vulva and an enlarged abdomen. On palpation, the spleen is larger than normal.
Laboratory Findings: The CBC is normal but the estrogen level is increased.

a) What is the likely diagnosis by the veterinarian?

   ____

b) Describe how the signs and history help confirm the diagnosis.

   ____

c) Explain the common treatment protocols for this problem.

____

d) What are the problems with asking pet stores for medical advice?

____

____

____

Case Study 3

History: An owner is concerned about her five-year-old neutered ferret. The ferret is pawing at its mouth constantly and the owner believes there is something stuck in the roof of its mouth. At times, the pawing becomes frantic. With further open-ended questions (questions that cannot be answered with yes or no) the owner also recalls episodes where the ferret seems weak, and walks with a stagger, "like it is drunk."
Physical Examination: There are muscle tremors and muscle wasting, but no foreign body is found in the mouth.
Laboratory findings: Blood glucose level is 65 mg/dl.

a) What is the veterinarian's diagnosis?

____

b) Describe how the signs and history confirm the diagnosis.

____

c) What is the likely outcome for the ferret?

____

Crossword Puzzle

Solve the following crossword puzzle relating to ferrets.

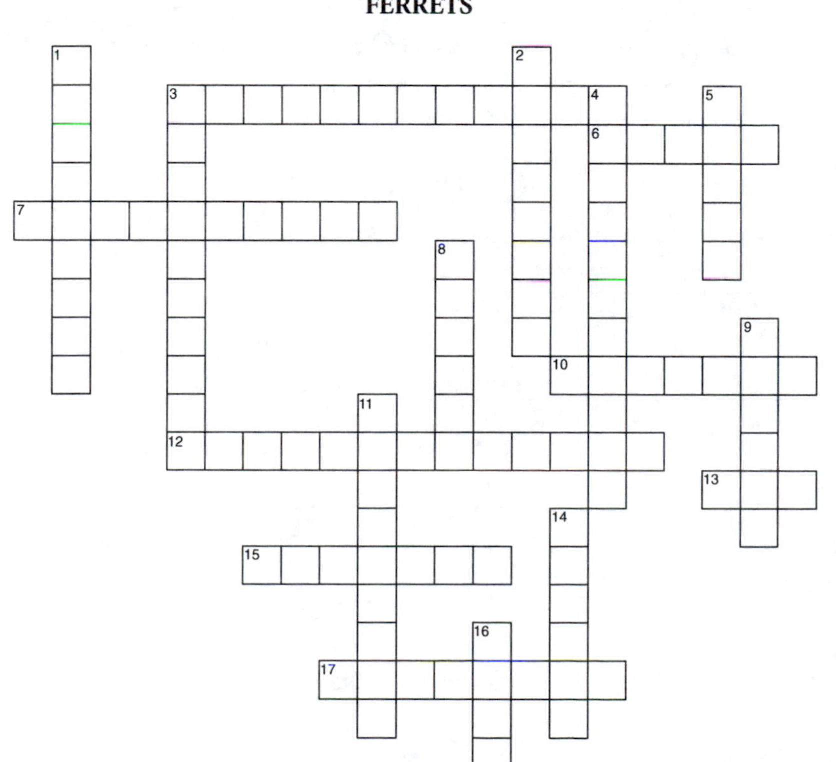

Across

3. recipe that never changes
6. non-steroidal anti-inflammatory drug
7. a condition from birth
10. animal capable of giving other animals disease
12. a condition of having reddened skin
13. male ferret
15. having to do with the stomach
17. itching, especially with inflammation

Down

1. disease affecting many animals in a specific location
2. first part of the small intestine
3. ad.lib.
4. exaggerated allergic response
5. polecat
8. blood in the stool
9. disease carrying insect
11. parasite transmitted by a mosquito
14. like decaying flesh
16. female ferret

Chapter 5

Rabbits

1. List the advantages and disadvantages of acquiring a pet rabbit, then decide how the rabbit is likely to meet your expectations as a companion animal. (List 4 of each.)

   Advantages:

   ____

   ____

   ____

   ____

   Disadvantages:

   ____

   ____

   ____

   ____

   Why would a rabbit make a good or not so good companion for you?

   ____

   ____

   ____

2. If the opportunity arises, visit a state or county fair or agricultural show. Discuss breed traits with the exhibitors and ask why they have chosen to breed and show rabbits. There will be a great variety of rabbits and an even greater number of reasons given. What are some of the difficulties exhibitors have experienced, what are the benefits?

   Difficulties:

   ____

   ____

   Benefits:

   ____

   ____

   ____

3. When you visit a pet store, note whether the rabbits and guinea pigs are housed together. Why should they be housed separately from one another?

   ____

4. Look at the environment carefully. Is the rabbit able to see ferrets and reptiles and large birds?

   Yes _____ No ____

Why is this important?

____

____

Case Studies

Case Study 1

History: An owner reports that her seven-year-old neutered male rabbit has had bloody urine for the last few days. The appetite is normal with no weight gain or loss, but the rabbit seems to sit "hunched over."
Physical Examination: Heart and respiratory rates are normal. CBC and Blood chemistries were declined, but the owner agreed to a radiograph. The radiograph below shows an abnormality in this rabbit.

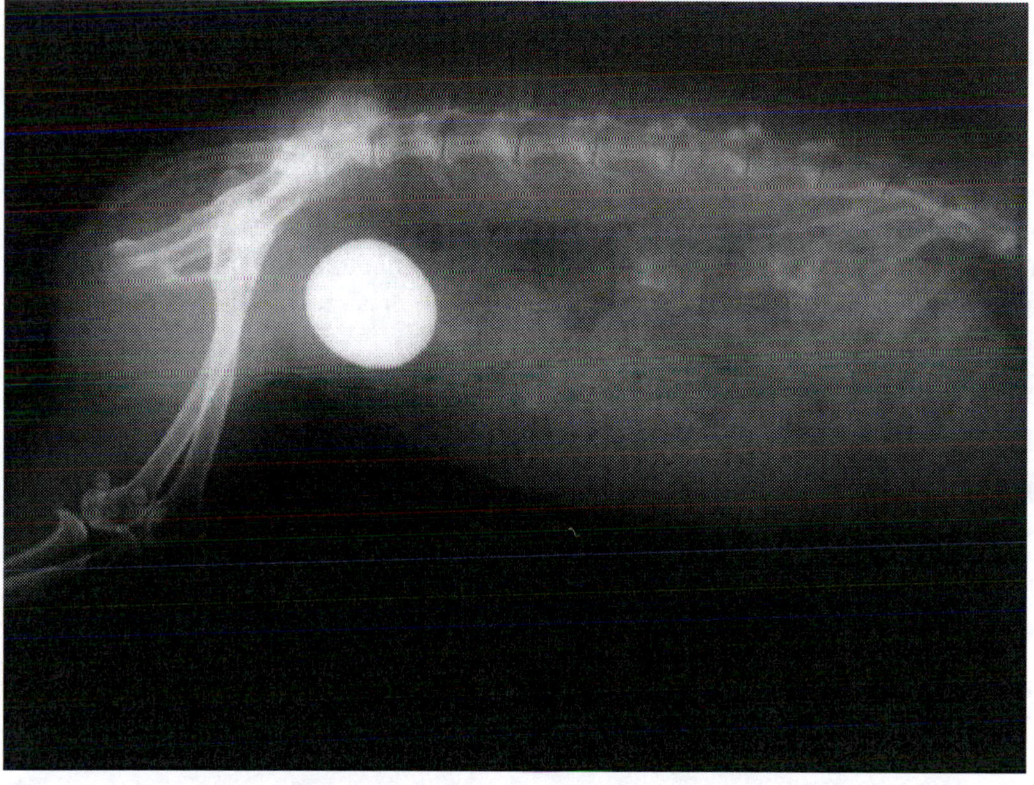

a) What can you tell from the radiograph about the abnormality?

____

b) How was the problem treated by the veterinarian?

____

c) What are the recommendations to prevent further problems like this?

____

____

Case Study 2

History: A four-year-old doe has become lethargic and weak. The owner has noticed a decreased appetite over the past four to five days and that very few droppings are being produced. What little fecal material there is is soft and watery. The doe sits with a hunched posture and is very tense.
Physical Examination: The veterinarian palpated the doe and noted a distended abdomen. The doe's temperature was below the normal range.

a) What could be the possible cause of these signs?

____

b) Describe how the signs and history would help to confirm a diagnosis.

____

c) What is the recommended treatment/prevention for this condition?

____

____

Case Study 3

History: A two year old buck has been sneezing and there is a thick mucous discharge around the nares. The owner reports that it has a decreased appetite. The rabbit is housed outdoors.
Physical Examination: There is dried mucous on the inside of the forepaws. The nasal passages are clogged with a yellowish discharge that is also evident around both eyes. The rabbit is slightly underweight and slow to respond. The footpads are heavily soiled with fecal material. The temperature is 104.5F, breathing is labored, but the heart rate is normal.

a) What is the most immediate medical problem that comes to mind?

____

b) How is this disease transmitted to other rabbits?

____

c) What can predispose a rabbit to this condition?

____

Crossword Puzzle

Solve the crossword relating to rabbits.

RABBITS

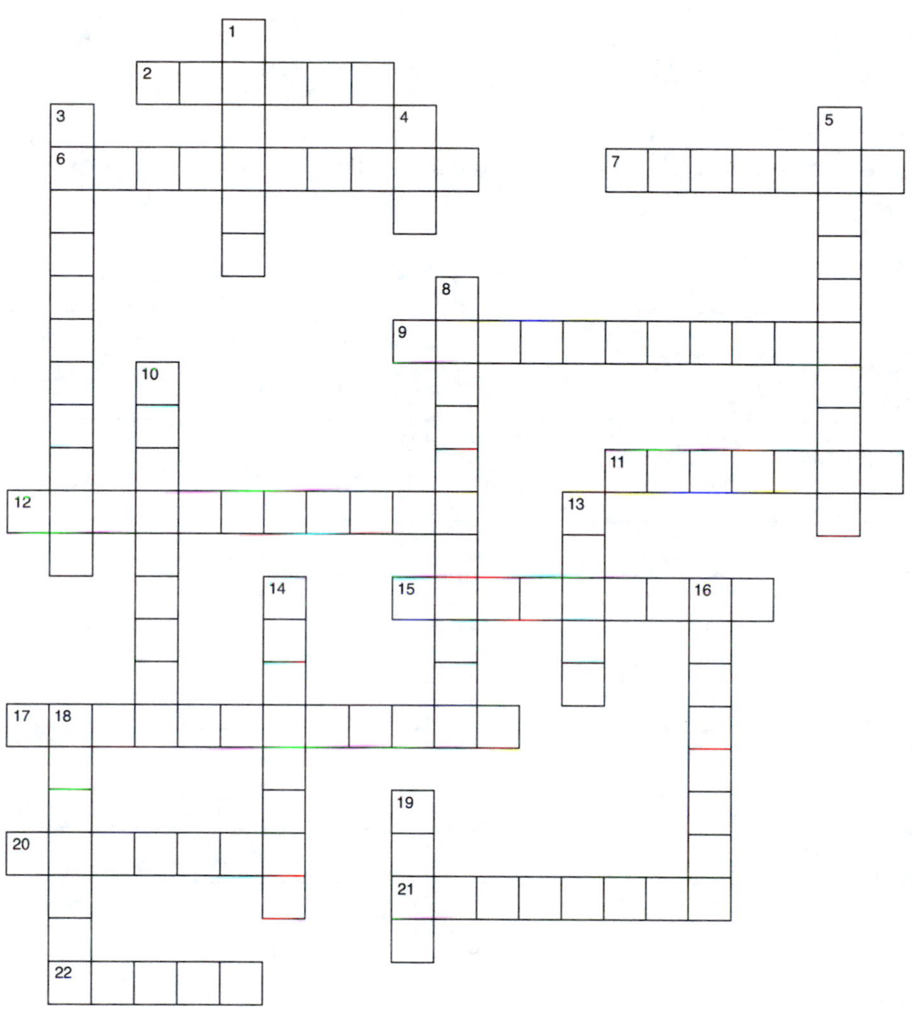

Across

2. being pregnant
6. inhaling fluids or solids into the lungs
7. difficulty breathing
9. another term for a hairball
11. a type of grass used for hay
12. another term for Wry Neck
15. the parasitic larva of certain flies
17. raising rabbits for meat
20. a legume grown to make hay and pelleted feed
21. have a blue color to the skin and mucous membranes
22. outdoor cage for a rabbit

Down

1. underground home for rabbit colony
3. process of giving birth
4. female rabbit
5. night feces produced by rabbits
8. active at dawn and dusk
10. production of milk
13. once domesticated animal that has established a populatin living in the wild
14. near the groin
16. inflammation of the nasal passage
18. a bladder or kidney stone
19. male rabbit

Chapter 6

Guinea Pigs

1. A friend has asked you for advice on choosing a pet. Would you recommend a rabbit or a guinea pig?

   Rabbit __________ Guinea pig ______________

2. List the advantages and disadvantages of each, compare what you know about them with what your friend expects from a pet.

   Advantages:

   Disadvantages:

3. Which one would you recommend and why?

4. What has caused the large area of blackness in the radiograph below of the guinea pig.

5. What factors contribute to this condition?

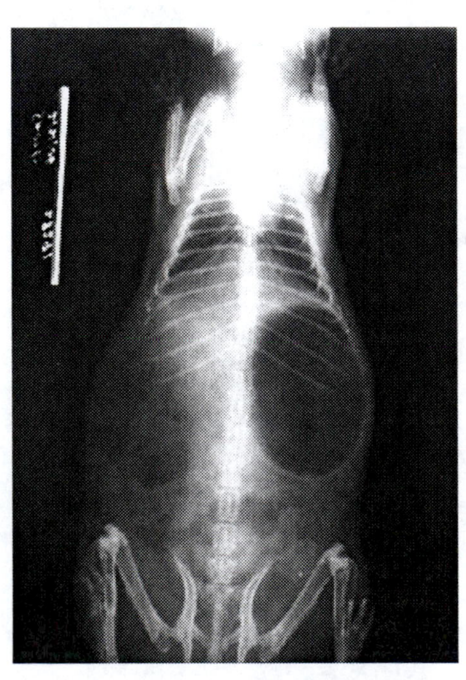

Matching

Match the following numbers with the appropriate letter definition:

1) drilling	_____	a)	a bad bite
2) dyspnea	_____	b)	Vitamin C
3) asorbic acid	_____	c)	tissue death
4) scurvy	_____	d)	disease caused by lack of Vit.C
5) dystocia	_____	e)	sound of warning or aggression
6) precocial	_____	f)	difficulty breathing
7) necrosis	_____	g)	difficult birth
8) malocclusion	_____	h)	born fully developed

Case Studies

Case Study 1

History: A two-year-old sow is lethargic and the owner hasn't seen her eating for several days. When the guinea pig attempts to move, she collapses and seems reluctant to get up again. The owner reports that she was rescued and lives with his-pet-rabbit. Both are fed rabbit pellets and some hay.
Physical Examination: The joints above both hocks are swollen. The patient has lost weight and has very little energy. Her mouth has small bleeding sores and the owner assumed that she got into something.

a) With a review of the history and diet offered to the guinea pig, what would be the most obvious reason for this condition?

____

b) How could this problem have been prevented?

____

Case Study 2

History: A three-year-old sow has been brought in by the owner, who noticed that her urine has become clear. The owner reports no other concerns, but casually mentions that the boar she acquired to keep her company is fine and they have been together for about two months.
Physical Exam: the sow is very weak and appears depressed. When persuaded to walk forward, she becomes ataxic. Her abdomen is enlarged on palpation, but is not painful.

a) What is the most likely cause of this condition?

____

b) What will happen if this condition is not treated?

____

c) What is the recommended treatment for this condition?

____

Case Study 3

History: A six-month-old intact boar, recently purchased from a pet fair, is presented with a yellow discharge from his eyes and nose. The owner says it is refusing food, hardly drinks any water, and is very lethargic. She can't take it back because the people who had all the bunnies and guinea pigs at the fair are gone.
Physical Examination: The temperature is below normal, and there is audible congestion when auscultating the lungs. The yellow discharge is thick and crusty over the nares and in the corners of both eyes. There seems to be some weight loss dehydration.

a) Considering the history, what would be the most likely cause of this condition?

____

b) How is this disease spread?

____

c) What clue was given in the history?

____

Crossword Puzzle

Complete the following crossword related to guinea pigs.

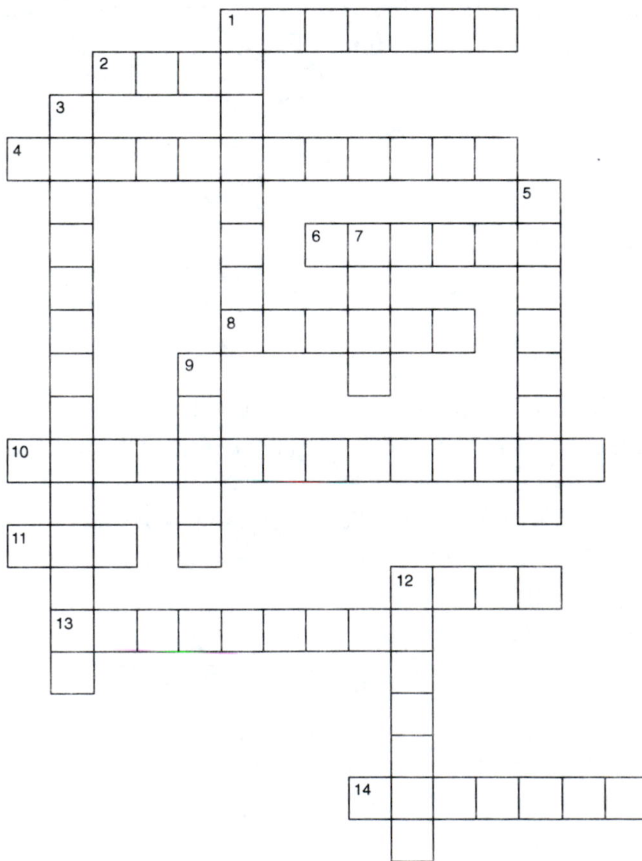

GUINEA PIGS

Across

1. difficulty breathing
2. male cavy
4. low blood sugar
6. disease caused by insufficient asorbic acid
8. another word pregnancy
10. another name for "bumble foot"
11. female cavy
12. country of origin
13. being born fully developed
14. name of a hairless guinea pig

Down

1. sound of warning
3. "porpupine-like"
5. difficulty giving birth
7. correct name for a guinea pig
9. bedding material that is toxic
12. bottom surface of foot

Chapter 7

Chinchillas

1. There is a great diversity of species to be found in South America. Print out a map of the South American continent and research the vast range of habitats (worldatlas.com). Place the various species in their ecosystems.

   How many of these overlap, or have adapted to varying conditions?

   ____

   ____

   What must be considered with these animals in captive environments?

   ____

   ____

2. Years ago, the exploitation of the chinchilla for its fur placed this species in extreme jeopardy in the wild. Access the www.CITES.org and view the current status of the chinchilla in the wild. How could the CITES listing be changed?

   ____

   ____

   ____

3. Write a researched but reflective paper regarding the status of the chinchilla and what might be done to restore a wild population. Consider not only the effects of the fur industry, but the popularity of chinchillas as companion animals.

   From your research, was Chapman instrumental in saving the chinchilla from extinction, or did he contribute to the loss of chinchillas in their natural habitat?

   ____

Case Studies

Case Study I

History: A six-year-old female chinchilla has been brought in because the owner noticed that the right eye is watering and appears to be inflamed. The discharge is clear but he reports that she is "squinting" during the day and thinks she may have scratched her eye. Otherwise, the patient appears to be fine.
Physical Exam: The eye is inflamed and partially closed with a clear discharge. Ophthalmic exam reveals no injury or foreign body.

a) What is the most likely cause of this condition?

   ____

b) What recommendations would be made to correct the problem?

   ____

c) What additional history would be needed for a diagnosis?

   ____

Case Study 2

History: A nine-year-old intact male chinchilla is presented because the owner has noticed it seems to have difficulty urinating and keeps chewing on its belly.
Physical Exam: On palpation, the bladder is full. The penis is visible and swollen. The chinchilla is unable to retract the penis.

a) What else is likely to be evident?

____

b) What is the cause of this problem?

____

c) How can this problem be corrected?

____

Case Study 3

History: A two-year-old male chinchilla has been brought in because the owner is concerned about a decrease in the droppings. What few fecal pellets there are seem normal and he reports that the chinchilla's appetite is good, or perhaps slightly less than normal, but the chinchilla just isn't very active. When questioned further, the owner reports that the diet consists of iceberg lettuce, free choice chinchilla pellets, and a raw carrot daily.
Physical Examination: Upon palpation, the veterinarian is able to determine that the intestines are impacted with feces. The chinchilla is obese and doesn't attempt to struggle.

a) What is a likely cause of the chinchilla's problem?

____

b) What are the ways to prevent this condition from occurring?

____

Crossword Puzzle

Complete the crossword related to chinchillas.

Across

1. area of hair release
6. outer cartilagenous part of the ear
8. inflammation of the conjuctiva
9. what needs to be provided to maintain the fur
12. little Chincha
13. non-productive attempt to vomit
15. defensive behavior
16. blockage of the esophagus

Down

2. business of raising chinchillas for fur
3. describes a coat pattern
4. abbrev, withholding of food
5. mountain range in South America
7. species included on CITES!
10. person who first brought chinchillas to the United States
11. area of dermis where hair grows
14. normal color for a chinchilla

Chapter 8

Hedgehogs

The African Pygmy Hedgehog was banned from importation because they were found to be carriers of anthrax.

1. What is anthrax?

   ____

2. What potential impact would an anthrax outbreak have on the agricultural/livestock industry?

   ____

   ____

   ____

   ____

3. Does anthrax have zoonotic potential? Yes __________ No __________
4. What is the protocol for handling an anthrax outbreak?

   ____

   ____

   ____

   ____

Key Search Words

Anthrax, African pygmy hedgehog, USDA/APHIS/Zoonotic diseases/ reportable disease, anthrax preventative measures and control

Case Studies

Case Study I

History: The patient is a four-year-old intact male hedgehog. The owner has noticed that there has been a drop in food consumption for the past week to the point where the hedgehog "won't even eat his mealworms." The hedgehog appears to be lethargic and is visibly thin.
Physical Examination: The hedgehog is ataxic and the temperature is subnormal. The veterinarian also notes ascites and a generally poor body condition. The hedgehog made no attempt to curl up during the exam.

a) What is a likely cause for these clinical signs?

   ____

   ____

b) Explain these signs and how they contribute to the diagnosis by the veterinarian?

   ____

   ____

   ____

c) What is the prognosis for this hedgehog?

Good __________ Poor ____________

Case Study 2

History: A year old female hedgehog is presented by the owner who states that "she seems to be dragging her hind legs." The sow has difficulty getting up and is reluctant or unable to curl into a ball.
Physical Examination: There is definite weakness in the hindquarters with a slow reflex response. The front limbs also exhibit a diminished reflex response. The hedgehog is alert, but ataxic, and unable to stand upright on all four legs.

a) What is the likely cause for this hedgehog's weakness and ataxia?

____

____

b) Explain the prognosis for this hedgehog.

____

____

c) If euthanasia is recommended, explain why.

____

____

____

Case Study 3

History: A four-month-old hedgehog is brought in by the young owner and his mother. The mother's chief concern is that the hedgehog, given to her son by a class mate, has rabies. The boy's response is, "But, Mom, I told you…." The mother's concern is that when the hedgehog is allowed out to wander around the living room at night, it starts foaming at the mouth and wiping the foam all over itself.
Physical Examination: Weight and vital signs are normal. The hedgehog is good health. It is very active and more interested in exploring its surroundings than curling into a ball.

a) What is the veterinarian likely to explain to the concerned mother?

____

____

b) What is the cause of this behavior in the hedgehog?

____

c) How is the visit to the veterinarian's office justified?

____

____

____

Crossword

Complete the crossword puzzle relating to hedgehogs.

Across

1. "road kill"
5. study of fossil records
7. forms hair, nails, spines, and horn
10. male hedgehog
12. invasive, spreading neoplasia
13. hedgehog behavior also called anting
15. accumulation of fluid in the abdomen
16. material on cage floor

Down

2. walk
3. skin condition
4. group of clinical signs suggestive of disease
6. a new growth, a tumor
8. young hedgehog
9. female hedgehog
11. "belly button"
14. female external genitalia

Chapter 9

Degus

1. As opposed to many exotic pets, degus are diurnal. What does this term mean?

 ____

2. Refer to your textbook; what is required for possession of degus?

 ____

3. To which government agency would you apply?

 ____

4. Contact the appropriate agency and request an application. Report back to your class with the details of the application and what is required. What is the greatest area of concern regarding degus?

 ____

Case Studies

Case Study 1

History: A four-year-old female degu is brought in by the owner. Her chief complaint is that the degu is losing the hair on both front legs. The degu's appetite and activity level are reported as normal. She feeds it guinea pig pellets and grass hay. The owner has only one degu, which she keeps in a 50 gallon aquarium. She uses pine shavings for bedding. The degu did have a rodent wheel, in fact about three, before the owner quit buying them because all the degu ever did was chew them up.
Physical Exam: There are areas of alopecia on both forelegs, it appears random and ragged with shorter hairs further up the legs. The veterinarian also notices a similar pattern in the hair coat of the lower abdomen.

a) What is the probable cause of the alopecia and disrupted hair coat?

 ____

b) Review what the owner says regarding the habitat. Incorporate your knowledge of degu behavior and suggest ways to correct and prevent this problem from developing further.

 ____

 ____

 ____

 ____

Case Study 2

History: A two-year-old male degu is brought in by the owner, concerned about recent weight loss. He also reports that the degu isn't as active as the others. It seems to want to eat, then just stops, pauses, like it hurts or something.
Physical Examination: Vital signs are normal, but the degu struggles when the veterinarian attempts to examine its mouth. He recommends a quick anesthetic mask down in order to examine the oral cavity. The owner agrees and the veterinarian reports back to the client.

a) What has the veterinarian been able to diagnose?

____

b) The veterinarian advises the client that he has been able to correct the problem. What has the veterinarian been able to do with the degu under anesthesia?

____

____

c) What further recommendations will be given to the owner to help prevent this problem in the future?

____

Case Study 3

History: A year old female degu is presented by the owner. Her concern is that the eyes appear to be clouding over and she says "I have the parents, and I think 'Mom' is blind." The other two degus from the same litter do not show any similar signs and are in apparent good health. The owner reports that she feeds her colony grass hay, rabbit pellets, fresh veggies, and dried fruit.

Physical Examination: The veterinarian performs an ophthalmic exam and confirms that the degu has early-stage, bi-lateral cataracts. She also requests a blood glucose level to which the owner consents, although not fully understanding the connection. As suspected by the veterinarian, the degu is hyperglycemic.

a) With the confirmation of cataracts, why did the veterinarian request the blood glucose level?

____

____

b) What is the connection between the cataracts and the glucose level?

____

____

c) Is this problem likely to be dietary, hereditary, or both? Explain your answer.

____

____

____

Crossword Puzzle

Complete the crossword puzzle relating to degus.

Across

3. a distinctive sound made by degus
5. Chilean squirrel or brush-tailed rat
7. complete loss of the skin on the tail
9. another term for a bad bite
13. feeding strictly on plant material
14. refers to figure of 8 shape of degu teeth
16. like a guinea pig

Down

1. number of toes on each degu foot
2. interpreting animal behavior by human feeling and emotions
4. fine power derived from volcanic ash
5. a common medical problem in degus
6. material that should never be used in a degue cage
8. abrasion of the rostrum
10. tooth color of a neonate degu
11. a type of grass hay
12. cage with different levels
15. legal requirement for keeping degus

Chapter 10

Hamsters and Gerbils

1. Hamsters and gerbils have long been considered "first pets." They are small, relatively inexpensive, and easy to care for. Which species would you recommend to a friend as a pet for his seven year old son?

   Hamster________ Gerbil ____________

   List the reasons for your choice.

   ____

   ____

   ____

2. What should an owner be aware of when completely cleaning the hamster habitat?

   ____

3. Hamsters are known carriers of Lymphocytic Choriomenengitis. How is this zoonotic disease transmitted?

   ____

Case Studies

Case Study I

History: A three-month-old Teddy Bear hamster is presented by the new owner. She reports that she has had the hamster for only three days and it is lethargic, has watery diarrhea and smells really bad.
Physical Examination: The hamster is dehydrated and feels cold to the touch. The hamster is reluctant to move, but when it does, ataxia is evident. There is a significant amount of watery diarrhea around the caudal area and it has a very noticeable odor. The veterinarian advises the client that the prognosis is grave.

a) What is the veterinarian's diagnosis?

   ____

b) What is the causative agent for this condition?

   ____

c) What are the contributing factors for this disease to suddenly appear?

   ____

   ____

d) Would the owner be wise to return the hamster to the pet store and have it replaced?

   Yes___________ No_______________

e) Why, or why not?

   ____

   ____

Case Study 2

History: A two-year-old male gerbil has been brought in because "it's starting to look like Rudolph, you know, the reindeer...." The gerbil is housed alone in a wire cage. It is provided with a water bottle and a food dish. The owner uses recycled paper bedding.
Physical Examination: The rostrum of the gerbil is inflamed and swollen. There is no hair around the nose.

a) Define this condition and it's probable cause.

It is a case of ______________________________

Caused by ______________________________

____

b) What suggestions can be given to the owner to prevent this condition?

____

____

Case Study 3

History: A two-year-old female hamster is presented because the owner's seven-year-old daughter has noticed that her hamster sleeps more than normal. It isn't eating very much and it is very difficult to rouse, even in the evening.
Physical Examination: The veterinarian has noted an abnormality in the abdomen. The hamster is underweight and lethargic.

a) What would be a likely diagnosis by the veterinarian?

____

b) What is the prognosis for this hamster?

____

c) Has the age of the hamster any influence on treatment options?

Yes______ No_____

Justify your response.

____

____

____

Case Study 4

History: A year-old male gerbil has been brought in by a concerned owner. The tail is completely de-gloved and there is evidence of dried blood. The gerbil is housed in an aquarium with a water bottle, hide box, and food dishes. The gerbil uses an open wire rodent wheel. The skin of the tail was found while cleaning the cage.
Physical Examination: The veterinarian confirms that the tail has been de-gloved and recommends amputation.

a) What does the term *de-gloved* mean?

____

b) Why is amputation recommended?

____

c) How could this have been prevented?

____

Crossword Puzzle

Complete the crossword puzzle relating to hamsters and gerbils.

Across

1. common medical condition of gerbils
4. the back of an animal
9. a tumor of lymphatic origin
10. severe bacterial disease of hamsters
11. the mating of closely related animals
13. an absence of hair
14. having one mate for life
15. inflammation of the lining of the nasal passages

Down

2. having an eye protrude from the socket
3. a small species of hamster
5. another name for a golden hamster
6. being born blind, hairless and helpless
7. the study of animals
8. oil producing gland in skin
12. non-colony animal

Chapter 11

Rats and Mice

1. The Whitten Effect was first recorded in mice. What are the potential benefits to agricultural animal breeders?

___
___
___

2. Why are mammary gland tumors so numerous, almost predictable, in domestic rats and mice?

___
___
___

3. Most mice and rats are masked down for anesthesia. What can you quickly make into a mask with readily available items found in the veterinary hospital?

___
___

Case Studies

Case Study I

History: A one-year-old intact male rat is brought in because the owner has seen him "crying red tears." She has also noticed that the rat is sneezing frequently and he doesn't seem to be eating as much.
Physical Examination: The veterinarian also notes some red crusting around the eyes and nares. The hair coat is poor and there are areas of red stain on the coat. With observation from a distance, he notes that the rat sits with a hunched posture and is slow to respond. There are also audible respiratory sounds.

a) What is the veterinarian's most likely diagnosis?

___

b) What is causing the "red tears?" (Fill in the blanks.)

It is ______________ from the ______________ gland.

c) What is the recommended treatment?

___

Case Study 2

History: A two-year-old female rat is brought in by her owner. The chief complaint is diarrhea and inactivity. The owner reports that the rat has been like this "for about 3 days...or slightly longer maybe."
Physical Examination: The veterinarian notes that the rat is very subdued and sitting hunched over in the carrying cage. It is lethargic and disinterested in its surroundings. When the veterinarian removes the rat from the cage, she notes that there is fecal material matted around the rectal area. The rat is also very dehydrated.

a) What is the probable diagnosis by the veterinarian?

___

b) What is the prognosis for this disease?

____

c) What is the causative agent of this disease?

____

Case Study 3

History: A two-year-old female rat is brought in because the owner noticed a small lump. The owner reports that it has increased in size from when she noticed it, about two weeks ago. There is another female in the cage and the cage mate appears healthy. At first, the lump didn't seem to bother the rat, but now, because of the increase in size and the location of the lump, the rat is favoring her right foreleg.

Physical Examination: The lump is located in the axillary region of the patient's right foreleg. The veterinarian notes that, due to the enlargement, the lump has interfered with movement of the foreleg but the foreleg itself does not seem directly involved. The veterinarian also notes that there is a small puncture wound on the top of the lump. The veterinarian has ruled out a mammary gland tumor.

a) What is her likely diagnosis?

____

b) What is the recommended treatment for this problem?

____

c) Describe the composition of the lump.

____

Word Scramble

1. INRPYHOPR _ _ _ _ _ _ _ _ _
2. HTICNILMTNEA _ _ _ _ _ _ _ _ _ _ _ _
3. NRUBZELEI _ _ _ _ _ _ _ _ _
4. LINPOITEORCE _ _ _ _ _ _ _ _ _ _ _ _
5. EIBEDATN _ _ _ _ _ _ _ _
6. IOONICGTN _ _ _ _ _ _ _ _ _

Define the unscrambled words

P ______________________________

A ______________________________

N ______________________________

P ______________________________

B ______________________________

C ______________________________

Crossword Puzzle

Complete the crossword puzzle relating to rats and mice.

RATS AND MICE

Across

8. common intestinal parasite of rodents
9. red substance produced by Harderian gland
10. Latin term that means "to gnaw"
12. having an ability to understand, be aware
14. early abortion of a litter to re-breed with a new male
15. a de-worming product
16. breed of rat with larger than normal ears

Down

1. term for color varieties in rats and mice
2. metric weight of approximately 2.2 pounds
3. when female mice begin to cycle at the same time
4. difficulty breathing
5. a condition of having a low body temperature
6. collective term for a group of mice
7. when hairs stand upright
11. a bacterial disease that affect the GI tract
13. "after death"

Chapter 12

Short-Tailed Opossums

1. Even though Short-tailed opossums (STOs) do not have a pouch, they are still marsupials. How are marsupials classified?

2. Instead of a pouch, how are the embryos of the short-tailed opossum nourished?

3. What one condition in the habitat would cause the ears of the opossum to become flaky and have a "crispy" appearance? (note: this is *not* the same as a condition referred to as "crispy ear").

Case Studies

Case Study 1

History: A three-year-old female short-tailed opossum with no previous health concerns has been brought in by the owner. She is concerned that the opossum is trying to "poop out" her intestines, as there is red, swollen tissue coming from the rectum.

Physical Examination: The veterinarian finds that the opossum does have a prolapsed rectum. The tissue has become very inflamed and swollen. No other diagnostics were performed or recommended.

a) What is the recommended treatment?

b) What are some of the causes of rectal prolapse in the Short-tailed opossum?

Case Study 2

History: A three-year-old intact male opossum is brought in for an exam. The owner has noticed that the bedding has become wetter than it usually is, and complains that she has to re-fill the water bottle every night, when normally the opossum doesn't need the water bottle changed or refilled for two to three days. Questioned further, the owner replies that there *might* have been some small spots that could have been blood.

Physical Examination: The veterinarian sees nothing significant during the initial physical exam and recommends a radiograph and urinalysis. The owner declines the radiograph due to finances, but agrees to a urinalysis.

Laboratory Findings from the Urinalysis: Significant amounts of both blood and pus in the urine sample.

a) What problem would the laboratory results indicate?

b) What is the recommended treatment for this problem?

c) If radiographs were taken, what, if anything, would help to confirm the veterinarian's diagnosis?

___

___

d) What husbandry concern arises from the owner's comments regarding the water bottle and the frequency with which it is being re-filled?

___

___

Case Study 3

History: A two-year-old female opossum has been brought in by the owner. He is concerned because he has noticed that the opossum has been losing some of the fur on the top of the rump. The opossum is eating and behaving normally and his concern is that the rodent wheel might be rubbing off the fur. He feeds her a diet of crickets, insectivore pellets, a small amount of cottage cheese with a few mealworms and a weekly pinky mouse. He also says that she "probably runs five miles a night" in her wheel and he has been reluctant to take it out of her cage "because she is so busy all night long."

Physical Examination: There is an area of alopecia on the rump that appears to be spreading to the lower lumbar region. The skin is visible, but there is no evidence of inflammation or flaky skin. The hairs which remain are not damaged in any way.

a) What is the suspected cause of this regional alopecia?

___

b) What is the recommended treatment?

___

c) Is the rodent wheel implicated in any way?

Yes ________ No ________

Crossword Puzzle

Complete the crossword puzzle relating to Short-tailed opossums.

SHORT-TAILED OPOSSUMS

Across

3. name for the young of many species
4. a neonate mouse or rat
6. an instrument that measures environmental humidity
7. part of the female breeding cycle, time of receptivity
9. mother of any animal
10. providing items that promote natural behaviors in captivity

Down

1. a young, lightly furred rat or mouse
2. eating both plants and animals
4. a grasping appendage
5. period of time between conception and birth
8. governing agency that issues permits to breed or possess certain species

Chapter 13

Sugar Gliders

1. There are many ethical concerns regarding the keeping of exotic animals as pets and the expectations people have regarding them. Discuss the concerns specific to the sugar glider; consider its natural habitat, social structure, and diet.

____
____
____
____
____
____

2. A friend has just dropped by and says to you, "guess what I found at the mall?" She rustles a cloth bag hanging around her neck and you hear what sounds like a miniature chain saw. How can you best help your friend understand the concerns of having a solitary sugar glider as a "pocket pet?"

____
____
____

3. What is the best way to educate your friend and help keep this glider healthy, both physically and mentally?

   Physical needs? ____
   ____
   ____

   Mental health needs? ____
   ____
   ____

Case Studies

Case Study I

History: A six-year-old female sugar glider is presented by a very upset owner because there is "something red" coming out of its rectum. The sugar glider is very stressed and vocalizes loudly.
Physical Exam: It is evident to the veterinarian that the rectum has prolapsed.

a) What other questions need to be asked regarding the history of this patient?

____
____
____

b) What are the contributing factors to a prolapsed rectum?

____
____
____

c) What treatment is the veterinarian likely to recommend?

___

___

___

Case Study 2

History: A five-year-old intact male sugar glider is presented by the owner. The chief complaint is that the glider seems weak. The owner also thinks it has lost some weight, but has no record of previous weight and is "just guessing" because it seems to be eating less.

Physical Exam: The glider is weighed with a gram scale and its weight is below the average for an adult male. Weakness with ataxia is also noted. The glider is reluctant to move and does not vocalize.

a) What further questions need to be asked of the owner?

___

___

___

b) How are these signs associated with the diagnosis?

___

___

___

c) What treatment is the veterinarian likely to recommend?

___

___

Case Study 3

History: A two-year-old intact male sugar glider is present by the owner. His main concern is that the glider "seems to get hurt in the cage all the time." He has carefully checked for the source of the injuries but cannot find any evidence of what might be causing the problem. He has brought the glider in today because there is a fresh laceration on the abdomen, approximately 1½ inches long. When questioned further, the owner replies that the glider cage is a two story condo with an "old, suspended ferret sack" for sleeping. He has added some natural wood branches. He has just this one sugar glider. There were two, but the other one died approximately six months earlier.

Physical Examination: The sugar glider has several old and fresh wounds on the thorax and abdomen. The right ear has been shredded and the left ear has a few healed puncture wounds.

a) What is the most likely cause of these injuries?

___

b) What is the most likely treatment for these injuries?

___

c) What further recommendations and advice can you offer to the owner?

___

___

___

Crossword Puzzle

Complete the crossword puzzle relating to sugar gliders.

SUGAR GLIDERS

Across

1. active during the night
4. lofty tree tops
6. connects the dam to the offspring during pregnancy
7. to glide
9. active during the day
12. searching for food
14. discarded food material

Down

2. having to do with trees, living in trees
3. newborn
5. the taste, or flavor
6. flap of fur covered skin
8. prefix for bone
10. spiders
11. a state of decreased activity
13. a young sugar glider or kangaroo

UNIT III

Chapter 14

Birds

1. In the boxes provided below, draw the beak and feet of a psittacine bird and the beak and feet of a passerine bird.

Beak	Psittacine	Feet
Beak	**Passerine**	**Feet**

2. What correlations can be made between the shape of a bird's beak and its diet?

____

____

____

3. Avian digestion is different compared to other species. Label the anatomical drawing of a bird's digestive tract and other organs.

Identify the following body parts of a bird.

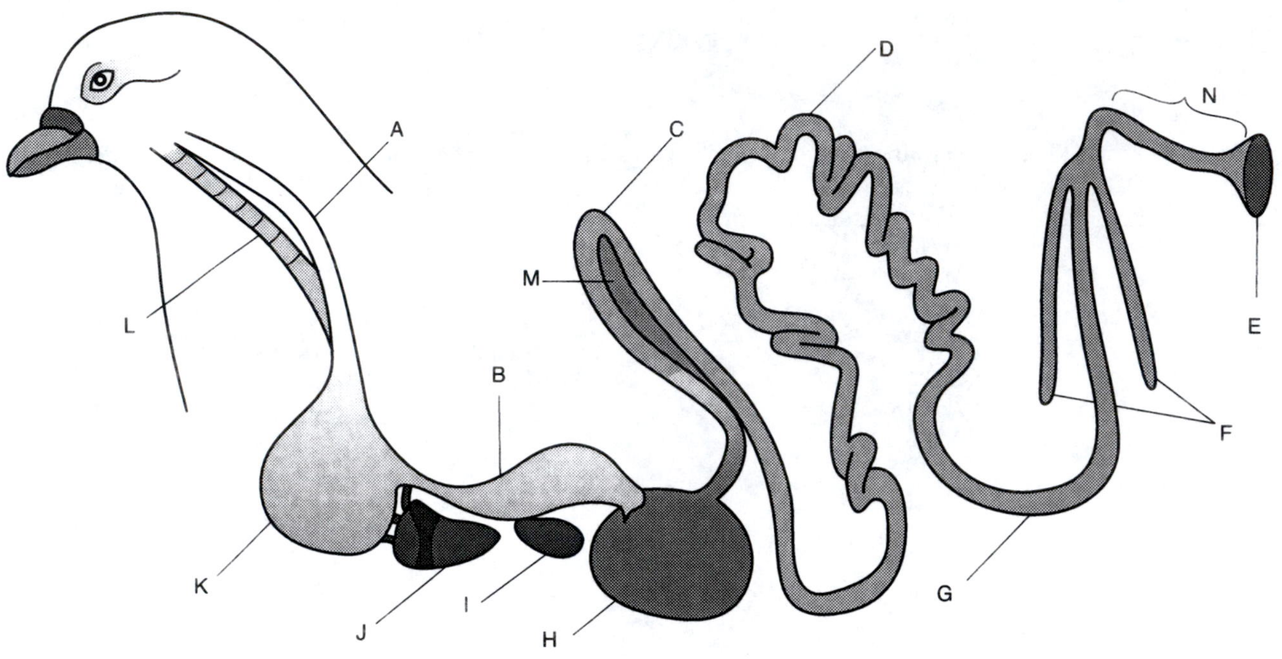

4. What are the three most distinctive differences between avian and other species?

____

____

____

5. List five food items that should never be offered to a bird.

____

____

____

____

____

6. What is specialized about the tongue of a lory?

____

____

7. What dietary needs must be met for lories?

____

Fill in the Blanks

In the following sentences, fill in the blanks:

a) Hunting with birds of prey is called ____________.

b) Pigeons are housed in a ____________.

c) Both male and female pigeons produce ____________ milk.

d) The ____________ are wing feathers and the ____________ are tail feathers.

e) When there are visual differences between the sexes, the species is referred to as being ________________________________.

f) Another name for the ventriculus is the ____________________.

g) A new and growing feather is referred to as a ____________________ ____________________.

h) Hookbills are psitticines, while finches are ____________________.

i) Pigeons and doves belong to a group called ____________________.

j) Down feathers provide ____________________ and powder down feathers provide ____________________ .

External Features of a Bird

Label the following external features of a bird.

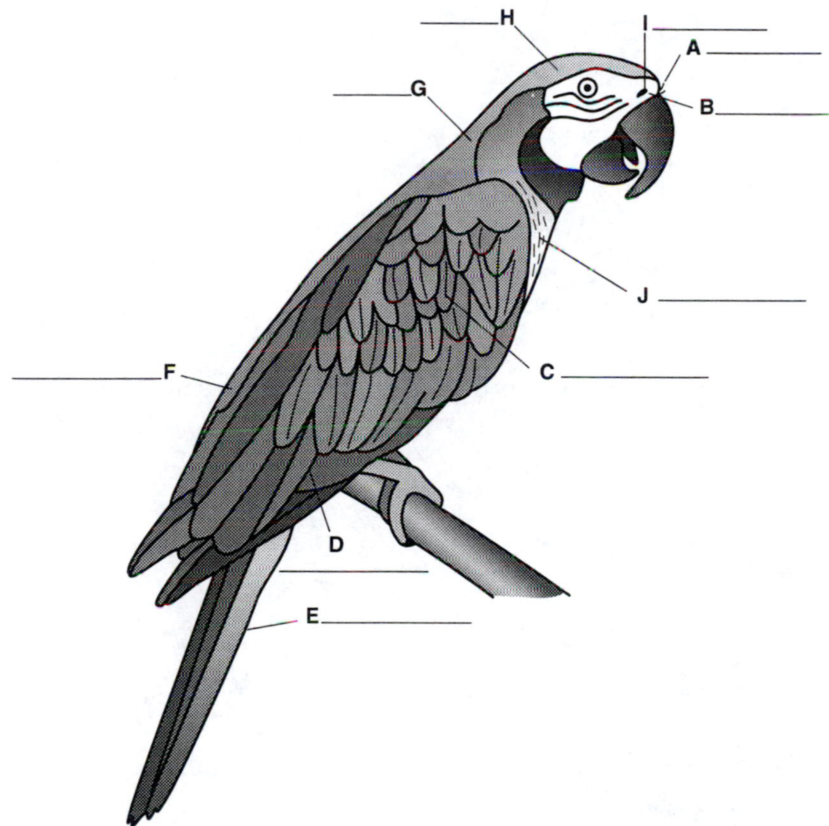

Restraint and Handling

1. Birds should never be restrained with pressure put on the front (ventral surface) of the bird. Explain why this is true and describe the correct method of avian restraint.

____

____

____

____

____

2. What is meant by the term "Step-up?"

____

3. Why is it important to keep a bird upright when it is being restrained?

____

____

4. When restraining for a jugular venipuncture, which side should be presented?

Left _______________ Right _______________

5. The urogenital system of birds is anatomically different from that of other species. Label the diagram of the urogenital system of a male bird. What is the significance. of one testicle being so much larger than the other?

____

Label the following organs of a male bird's urogenital area

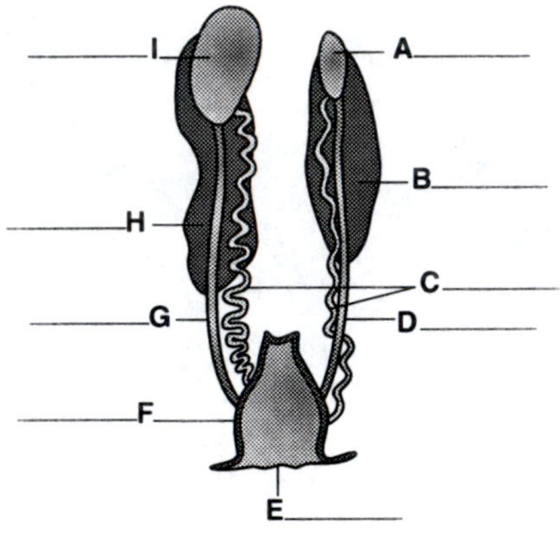

6. The respiratory system of birds is unique as it also assists in flight. Label the following diagram of the birds respiratory system.

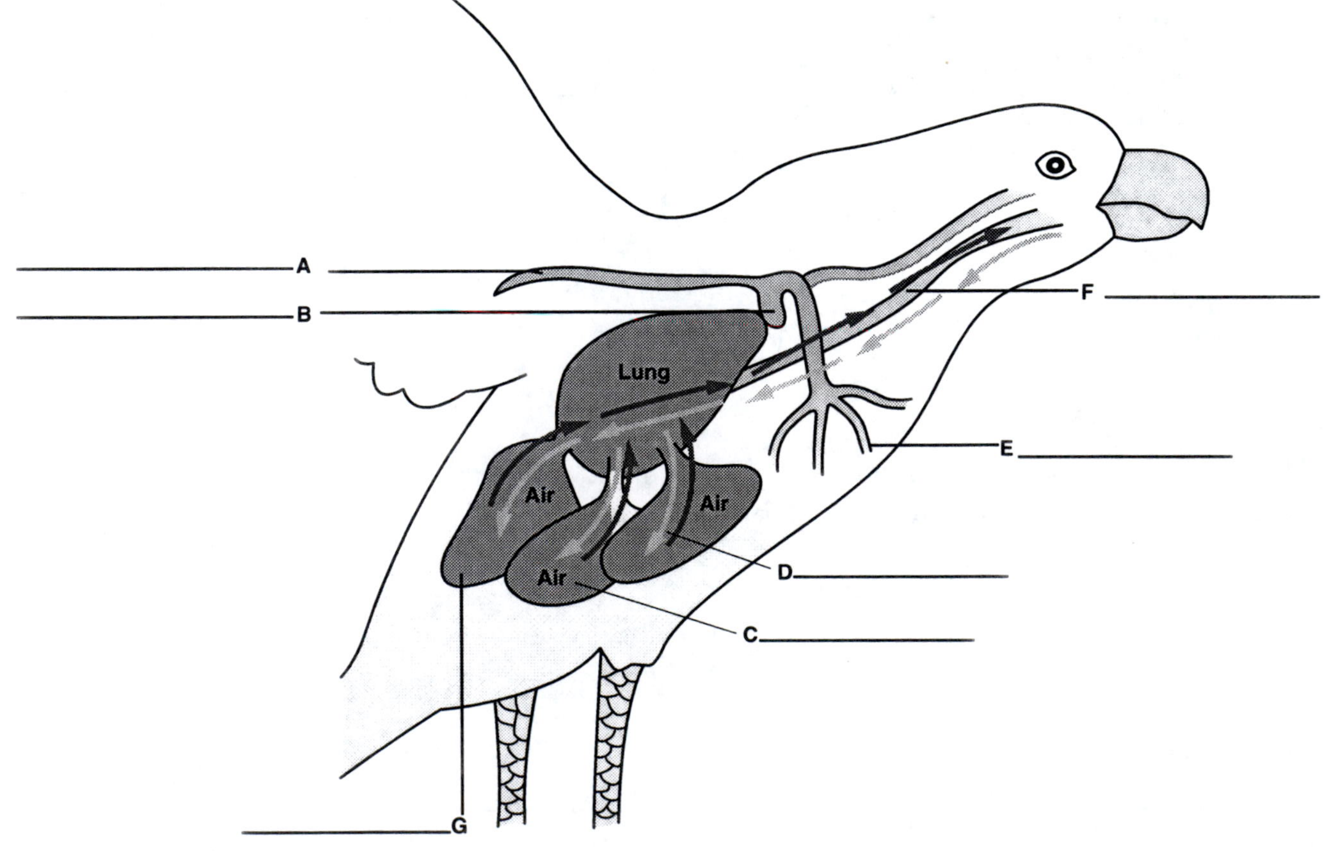

7. Explain the route of air transfer in birds from inspiration to expiration.

____

____

____

____

Medical Concerns

1. Birds can transmit disease to humans. Which avian disease is called *psittacosis* in humans?

____

2. What vaccination is available for companion birds to protect them from a fatal disease which causes "many tumors?"

____

3. What is the difference between the choana and the cloaca?

   Choana ______________________ Cloaca ______________________

4. What are two advantages of using inhalant anesthesia in birds?

____

____

Case Studies

Case Study 1

History: The owner of an eight-year-old Umbrella Cockatoo has been presented for sedation because it has become aggressive toward family members. This behavior started "about a month ago," and has become such a problem that it is difficult for anyone to handle the bird without being bitten. Further questioning determines the following from the owner: "We bought this huge new cage, with a playpen on top and everything. It was recommended by the breeder and we thought it would give her more room to play . . . she likes to be out of her cage . . . it's huge, higher than my head."

Physical Examination: The bird appears to be in excellent health. It is very interested in the examination. It willingly *steps-up* to the technician from the table perch. The feathers are immaculate and the bird seems to enjoy "showing off" for the clinic staff. The owner is astonished: "She would *kill us* if we tried that!"

a) Why has the bird become aggressive?

____

____

b) Why would sedation not be an appropriate approach to this problem?

____

____

c) Suggest methods that should be used to re-establish a healthy, happy relationship between the bird and its "human flock."

____

____

____

____

Case Study 2

History: A twelve-year-old Meyer's parrot is presented by a concerned owner. He has noticed that during this last molt, the new feathers look *weird* and some of them are a different color. He is also concerned that the beak is growing at an odd angle and has small black spots that were not there before. The owner also comments that he enjoys attending bird fairs and visiting all the pet stores, *just to look*, but he hasn't seen anything like this in other birds.

Physical Examination: There are several small fractures along the length of the beak. The beak has become malformed and appears to be twisting. The new feathers are not growing straight from the feather shaft, but curl and appear *clubbed*. Several of the new feather shafts are necrotic.

a) Given all the relevant history, what is the likely diagnosis by the veterinarian?

____

b) What is the prognosis for this bird?

____

c) What extra precautions need to be taken by those in contact with the patient and in cleaning the exam room?

____

____

____

____

d) What are the methods of transmission for this disease?

____

____

____

Case Study 3

History: A 15-year-old Blue & Gold Macaw is brought in by the owner for "some kind of test." The owner states that his physician said to "take the bird to the vet." The owner has been ill with flu-like symptoms for the past month. While taking the history, the staff member notices the close bond between the macaw and its elderly owner, who exchanges several *kisses* with the bird as he comforts it in the strange surroundings. The owner comments sadly, "He's my best buddy, all I have since my wife died a couple of months back. It was just us, the three of us."

Physical Examination: The bird appears to be healthy and has a bright attitude. The feathers and beak are normal and the weight is within the normal range for a Blue & Gold Macaw. The bird's droppings are normal.

a) What disease does the physician suspect this bird may have?

____

b) Discuss the connection to the owner's heath and his beloved bird. Consider all the history, observations by the staff, and comments from the owner.

____

____

____

____

____

c) What tests will be given to determine if the physician's suspicions are correct?

____

____

d) If the macaw tests positive, how will this disease be treated for both the human and avian patient?

___

e) Which groups of people are more susceptible to this zoonotic disease?

___ ___ ___

Crossword Puzzle

Complete the crossword puzzle relating to birds.

BIRDS

Across

1. the "true stomach" in birds
2. expandable food storage pouch
4. location of bristle feathers
6. a hookbill
10. small, hair-like feathers
14. the keeping and breeding of birds
15. the correct name for the gizzard

Down

1. grooming the feathers
2. causes psittacosis in humans
3. wings feathers
5. feathers that provide warmth
7. oil secreting gland at the base of the tail
8. literally, "many tumors", contageous disease of birds
9. tail feathers
11. perching bird
12. name for an immature pigeon
13. the fleshy part at the top of the beak

Unit IV

Chapter 15

Reptiles

1. Reptiles are episodic breathers. What does this mean?

2. If a snake does not have a healthy ecdysis, what happens and what can be done to correct this problem?

3. What common injuries can be caused by inappropriate cage furnishings?

4. What are the clinical signs of a reptile with an upper respiratory tract infection?

5. Explain how to auscultate a lizard.

6. How should water be provided in a chameleon cage and why is this necessary?

7. Name one common site for blood collection in each reptile species.

   a) Tortoise: __________

   b) Snake: __________

   c) Lizard: __________

Anatomical Diagrams

1. Label the internal organs in the diagram of lizard anatomy. Is this lizard a male or a female? Male ____________
   Female ____________

Ventral view of a female lizard

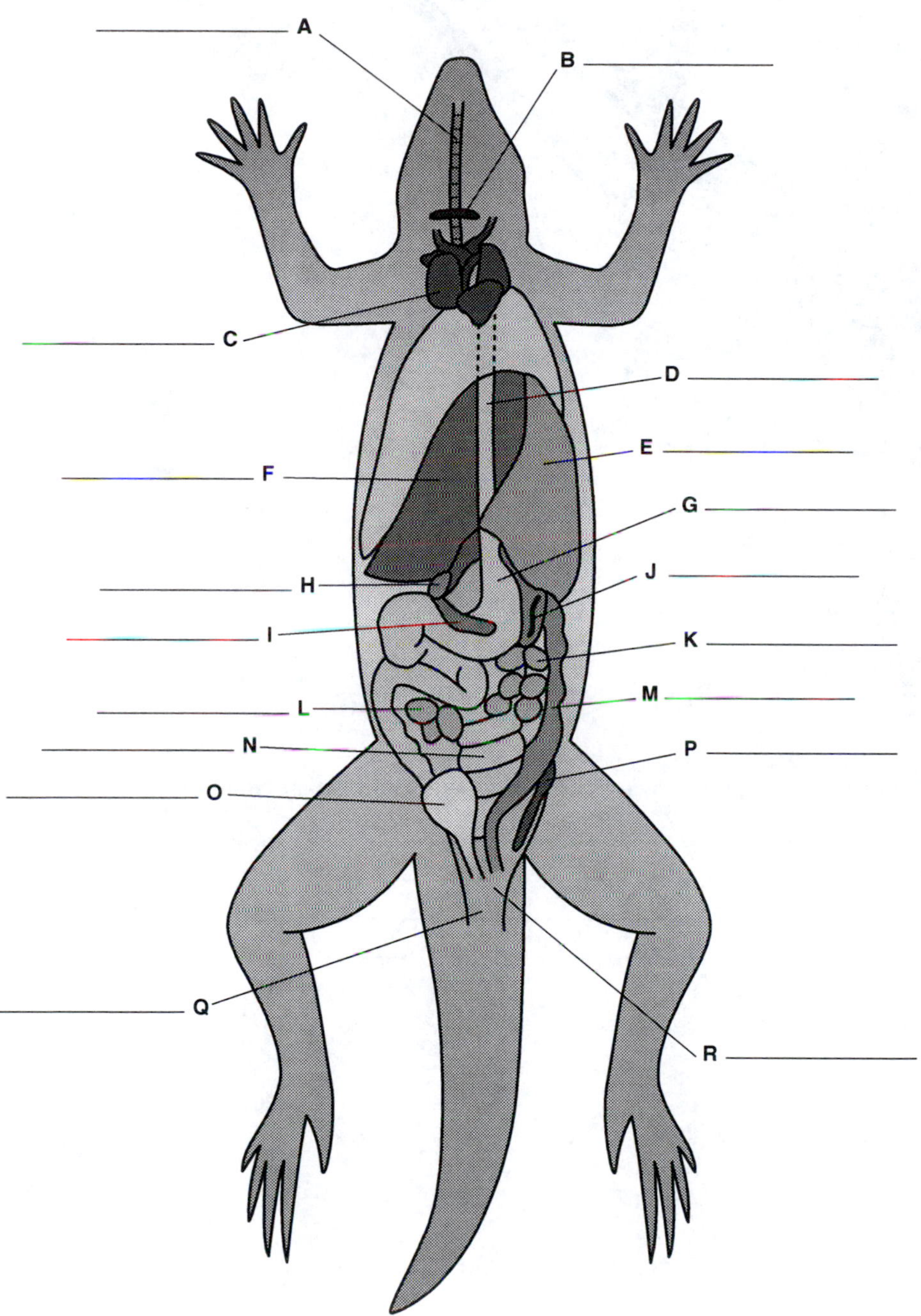

2. The reptile heart is different from the mammalian heart in that it has (finish the sentence)

____

3. Label the area of a reptilian heart in the diagram.

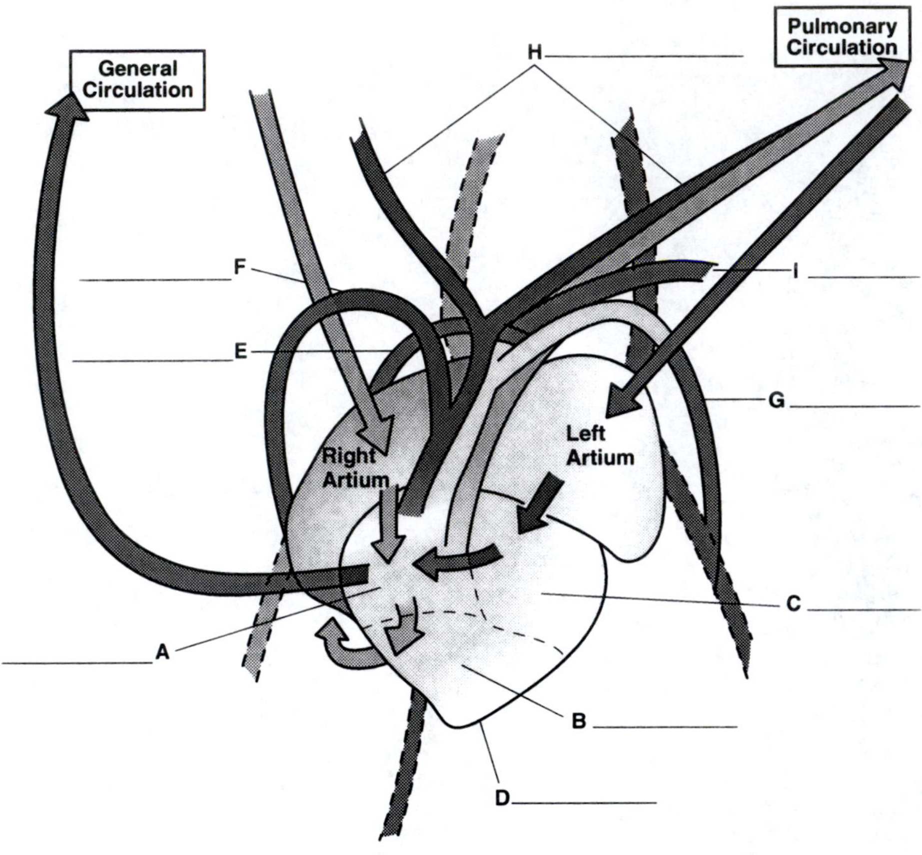

4. Describe the flow of oxygenated blood through the heart, beginning with pulmonary circulation.

___

___

___

___

5. The internal organs of chelonians are, in part, directly connected to the carapace. Label the internal organs of the male turtle.

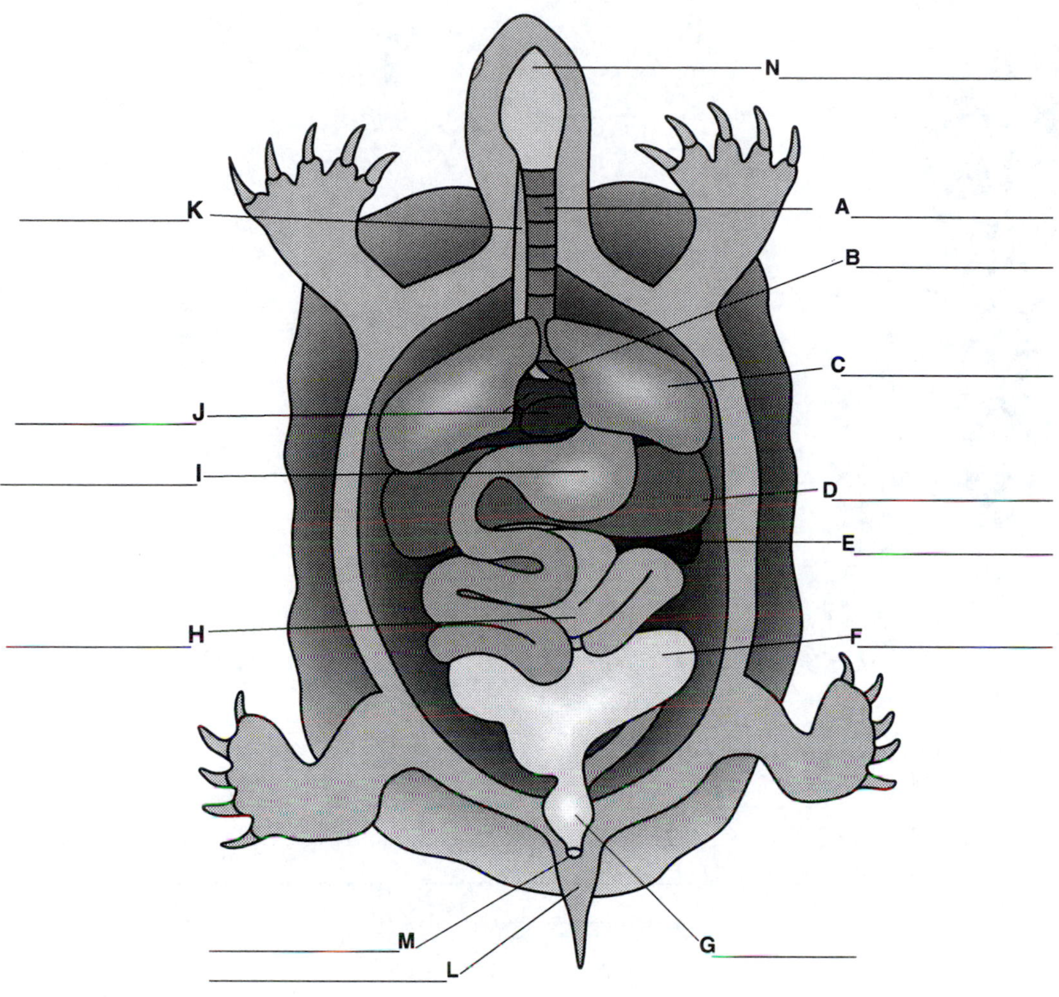

6. It is important to understand the renal portal system of reptiles because it affects drug delivery. If administering an injectable drug, which injection sites should be used?

____

Label the drawing of a lizard's renal portal system.

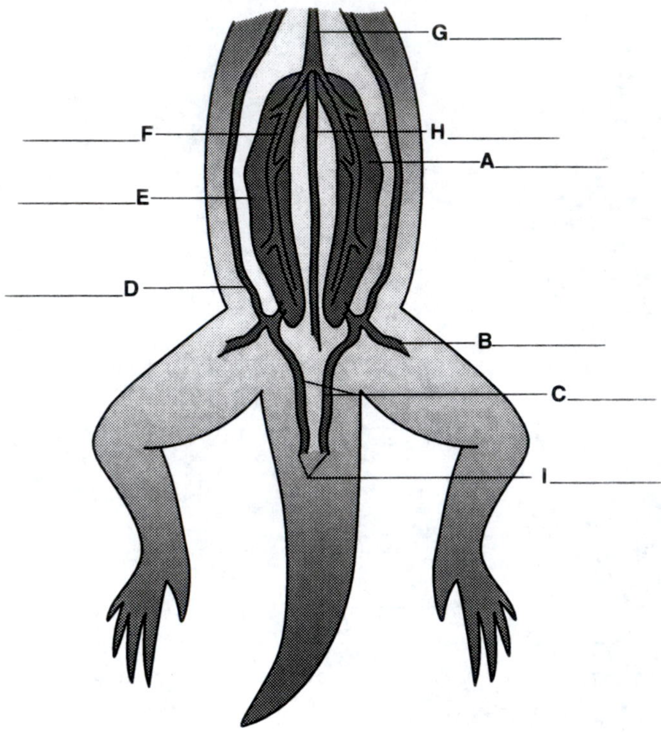

7. Explain blood flow through the renal portal system.

Case Studies

Case Study I

History: A juvenile green iguana has been presented by the owner because it is not eating and has difficulty moving. The owner purchased it two weeks earlier from a local pet store and houses it in a ten gallon aquarium.
Physical Examination: The left rear leg is swollen and the veterinarian suspects a fracture of the femur. The iguana is lethargic and rests on the exam table with its eyes closed. The veterinarian also notes that the iguana's pelvic bones are prominent and there is little muscle mass in either hind limb. The veterinarian also believes this to be a spontaneous fracture.

a) What disease condition could have caused the spontaneous fracture?

b) What other questions should the owner be asked regarding husbandry?

c) In assisting this client, what would you advise regarding

Diet ______

Lighting ______

POTZ ______

d) How is this condition treated?

______

Case Study 2

History: A seven-year-old Ball Python is presented by the owner. The chief complaint is that the snake is refusing to eat. The owner has had the snake for two years and reports that it has been healthy and eating regularly. He tried feeding the snake a week ago and the prey was refused. He thinks the snake may be "trying to hibernate" because it yawns a lot.

Physical Examination: The snake is having difficulty breathing and "yawns" or gapes during the veterinarian's exam.

a) What is the most likely problem with this snake?

______

b) What other major signs did the snake exhibit for the veterinarian to confirm her diagnosis?

______

______

______

c) What are the veterinarian's treatment recommendations?

______

______

Case Study 3

History: An adult green iguana is presented by the owner who is concerned because "he seems sick." She reports that the iguana is not eating, and has become "tamer." It hasn't produced anything, stool or urates, for days. She believes it to be a male because it has been aggressive and turns "kind of gold, or red sometimes." The iguana usually eats well. The diet consists of kale, mustard greens, broccoli, and zucchini.

Physical Examination: The veterinarian notes that the iguana is depressed and has a swollen abdomen. He recommends radiographs and the owner agrees. The radiographs clearly show multiple round masses in the abdomen as well as impacted fecal material.

a) By reading the radiograph the veterinarian is able to confirm his diagnosis. What is the problem?

______

b) The owner is very surprised by the diagnosis. Does color change and/or aggressive behavior indicate the sex of an iguana?

Yes ______ No ______

c) What is the recommended treatment?

______

Case Study 4

History: A medium sized Red-eared Slider is presented by a reptile rescue organization. It has an unknown history, except that the previous owners bought it when it was "very small," two years earlier. The volunteer is concerned about its health in general as he has noticed that some of the scutes of the shell are missing.
Physical Examination: There are several scutes missing and the plastron has areas of redness. The skin of all four legs is flaking and erythematous.

a) What is the veterinarian's diagnosis?

____

b) What can predispose a turtle to this problem?

____

____

c) What is the recommended treatment for this condition?

____

____

Case Study 5

History: A nine foot Columbian Red-Tailed Boa is presented by the owner, who is concerned that something is "driving it nuts." When questioned further, he reports that he hasn't had the snake for very long. It was in a wooden crate in a pet store and now he has noticed "things" moving under the scales.
Physical Examination: The veterinarian's exam confirms that the boa has a severe infestation of snake mites, *Ophyionyssus natricis*.

a) What are other signs that could indicate that a snake is infested with mites?

____

____

____

b) Explain the life cycle of the snake mite.

____

____

____

____

c) What is the recommended treatment for this snake?

____

____

____

Crossword Puzzle

Complete the crossword puzzle relating to reptiles.

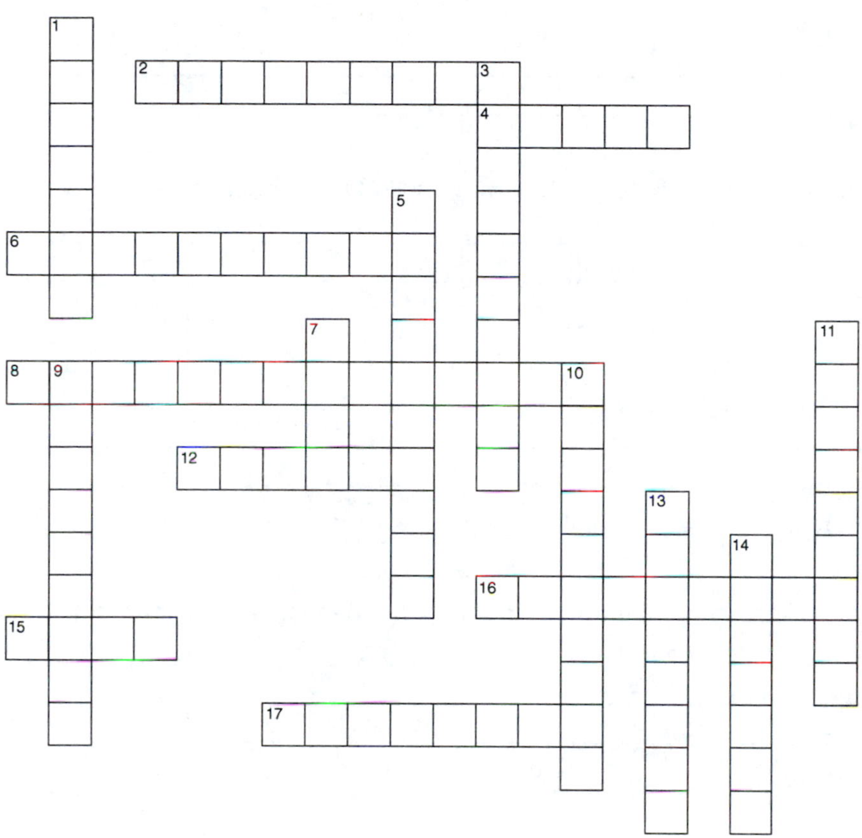

Across

2. the highly developed organ in a snake, near the tongue, that picks up a scent
4. number of chambers in a reptile heart
6. to listen with a stethoscope
8. having teeth that are continually reabsorbed and replaced
12. group of reptile eggs
15. optimum temperature range
16. reptile hearts have only one
17. lower part of a chelonian shell

Down

1. "cheezy," especially pus
3. another name for mouth rot
5. species that seek sunlight for warmth
7. disease caused by excess urates that are deposited in the joints and intestinal tract
9. egg-laying
10. group that includes turtles and tortoises
11. dependant on external temperature
13. upper part of a chelonian shell
14. when a reptile sheds its skin

UNIT V

Chapter 16

Amphibians

1. What are the three general types of amphibians?

2. What are some typical signs of an amphibian that is not housed within the POTZ for the species?

3. Explain the life cycle of amphibians.

4. What precaution should be taken before handling any amphibian?

5. What are the reasons for the handler to take such precautions, both for the handler and for the amphibian?

Case Studies

Case Study 1

History: An adult leopard frog of uncertain age has been presented by the owner because the frog is not eating. She has also noticed that there appear to be sores on the feet.
Physical Examination: The veterinarian determines that there are ulcers on the legs and toes, with inflammation and edema of the limbs.

a) What is the likely cause of the veterinarian's findings?

b) What is the recommended treatment for this condition?

Case Study 2

History: A juvenile White's Tree Frog has been brought in by the owner. The froglet was purchased two months earlier from a local pet store. The frog has grown considerably and the owner reports that it eats well. Her concern is that she has noticed two small bumps on one hind leg that were not evident before.
Physical Examination: Closer examination by the veterinarian reveals two obvious nodules on the right rear leg, and another on the left leg. The nodules appear to be isolated and of approximate size.

a) What has caused these nodules?

b) What is the recommended treatment?

c) Discuss the likely source (wild caught or captive bred) and the probability that all of the frogs in the same shipment will have this condition.

Case Study 3

History: An amphibian is presented to the clinic staff for identification. It was bought some months earlier as a "tadpole," but now has developed an elongated body, four legs, and a tail. It is approximately three inches long. It is definitely not the frog she was expecting but she is still very interested in keeping it and learning more about what she has.
Physical Examination: The amphibian is in apparent good health. It appears to be mature as there is no evidence of gills. The staff refers to field reference guides. They determined that it is a small salamander, but the species is not positively identified.

a) What general information could be given to the client regarding the husbandry of her salamander?

Crossword Puzzle

Complete the crossword puzzle relating to amphibians.

AMPHIBIANS

Across

1. amphibians with tails
4. living on land
6. changing body shape through life stages
7. newly hatched fish
10. group that includes frogs and toads
11. to lay masses of eggs
12. the study of reptile and amphibians
13. clasping the female to fertilize expelled eggs

Down

2. living in the water
3. retaining infantile characteristics as an adult
5. bacteria that causes Red Leg
8. a habitat that includes plants and animals
9. the immature stage of frogs and toads

UNIT VI

Invertebrates

Invertebrates are species that have an exoskeleton, rather than an endoskeleton. Word prefixes are essential in determining word meaning.

1. Define the prefix *endo*.

 ____

2. Define the prefix *exo*.

 ____

3. List six other words that begin with either *endo* or *exo* and define them. For example,

 Word: *endoscopy* Definition: to scope inside
 Word: *Exophthalmia* Definition: protrusion of the eyeball, the eye is out of the socket
 Word: ________ Definition: ________
 Word: ________ Definition: ________
 Word: ________ Definition: ________
 Word: ________ Definition: ________
 Word: ________ Definition: ________
 Word: ________ Definition: ________

4. The most popular and readily available scorpion is the Emperor Scorpion. What is the natural habitat of the Emperor scorpion?

 ____

5. Both scorpions and tarantulas are known as *liquid feeders*. What does this mean?

 ____

6. What is the reason that all species of Hermit Crabs are wild caught and cannot be captive bred?

 ____

7. A friend of yours has just acquired some hermit crabs. She tells you that they make "these weird noises" during the night. What is the term you would use to define this sound?

 ____

8. How is this sound produced?

 ____

Chapter 17

Scorpions

1. What effect does the venom from a scorpion sting have on its prey?

____

2. Explain how scorpions feed on their prey.

____

____

____

3. How is the sex of a scorpion determined?

____

____

Chapter 18

Tarantulas

1. What are the two types of tarantula venom and what effect does each have on the body?

   Types: ______________ ______________

   Effect: ______________ ______________

2. What is the purpose of the urticaric hairs on the tarantulas body?

____

3. Describe the appearance and behavior of a tarantula that is beginning to molt.

____

____

4. Describe the appearance of a tarantula that is sick or dying.

____

Chapter 19

Hermit Crabs

1. What is the name and purpose of the large claw of the Hermit Crab?

   Name: ____________________

   Purpose: ____________________

2. When does cannibalism most often occur in Hermit Crabs?

   ____

3. Describe the ideal habitat for a Hermit Crab.

   ____

   ____

   ____

   ____

Case Studies

Case Study 1

History: An owner telephones concerned that his tarantula is dead. It did not eat anything during the previous week, and now it has been found completely upside down in the habitat, with all of its legs "in the air."

Physical Examination: There is none, as the owner refuses to bring in a "dead tarantula" for an examination.

a) What would be your response to this type of telephone call?

   ____

   ____

   ____

   ____

b) What is the reason the tarantula has stopped feeding and is in this very strange position?

   ____

c) If the tarantula was not eating due to illness, how would the body position be different?

   ____

Case Study 2

History: A concerned owner presents an Emperor Scorpion, concerned that it has "maggots." She states that, three days ago, it was covered in "little white, wiggling things." Now, they have all disappeared. She is concerned that her home may have been infested with "whatever those things were."

Physical Examination: The veterinarian's examination determines that this is a healthy, robust scorpion. He asks two questions of the client: "Where were these *things* on the scorpion?" and "What did they look like?" The client replies that they "were on the back of the scorpion, they were sort of white, or pale, and had black dots at one end."

a) Considering the additional, open-ended questions asked by the veterinarian, what can you deduce from this additional information?

   ____

   ____

   ____

b) What is the likely explanation for the disappearance of these "things?"

____

____

Crossword Puzzle

Complete the crossword puzzle relating to invertebrates.

INVERTEBRATES

Across

4. silk producing organs
5. the "blood" of a tarantula
9. sound produced when two body parts are rubbed together
12. group of animals with 4 pairs of legs
13. animals with jointed legs
14. the mountparts of a scorpion
15. external hard shell made of keratin

Down

1. the end of a scorpion's tail
2. joined head and thorax
3. accordian pleated tissue where gas exchange occurs in tarantulas
4. an immature tarantula
6. ability to deliberately release a limb
7. animal without a backbone
8. largest claw of a hermit crab
9. a sperm packet
10. feathery looking appendages behind each pair of scorpion legs
11. an animal with ten legs
13. the stinger of a scorpion

UNIT VII

Chapter 20

Alpacas and Llamas

1. Alpacas and llamas are members of the camelid family. Camelids are divided into Old World and New World Species. What are the two Old World Species?

    ______________________ ______________________

2. What are the differences between Llamas and alpacas?

    Llamas: ______________________

    Alpacas: ______________________

3. What type of diet is recommended for alpacas?

    ____

    ____

4. There are four species of South American camelids. The alpaca and llama are not found in the wild. What are the other two species?

    ______________________ ______________________

5. What vaccinations are administered to alpacas?

    ____

    ____

6. Explain the differences between the stomach of an alpaca and that of a cow.

    ____

    ____

    ____

    ____

7. What are the differences between the suri and the huacaya?

    ____

    ____

Case Studies

Case Study I

History: In August, a client requests a farm call regarding her pet alpaca. The owner reports that it won't get up.
Physical Examination: The veterinarian examines the fleece, which is very long and matted. It has not been shorn since the owner bought the alpaca last year, as she prefers the "shaggy look." A black tarp has been tied between two tree branches to provide shade, but at 4:00 p.m. the tarp offers no protection from the sun. While taking a rectal temperature, the veterinarian comments to the owner regarding how hot the summer has been, and that it must still be more than 90°. The veterinarian records the rectal temperature at 105.8°. He then asks the owner if there is a garden hose nearby.

a) What is the diagnosis?

    ____

b) Why has the veterinarian asked for the hose?

____

c) What recommendations will he give to the owner?

____

____

Case Study 2

History: A client brings in a 20-week-old cria. His concern is that it may have a fractured leg.
Physical Examination: The veterinarian palpates the leg and determines there is no need for a radiograph as he detects no fracture. He also palpates the other three legs and carefully notes their comparative length and density, gently flexes each joint, then palpates the length of the spine. The owner is greatly surprised when the veterinarian asks about the owner's de-worming schedule. The owner said he hadn't done it yet, but intended to "this fall."

a) What does the veterinarian suspect is the problem with this cria?

____

____

b) What is the connection between the de-worming program and the lameness in the cria?

____

____

Crossword Puzzle

Complete the crossword puzzle relating to camelids.

ALPACAS AND LLAMAS

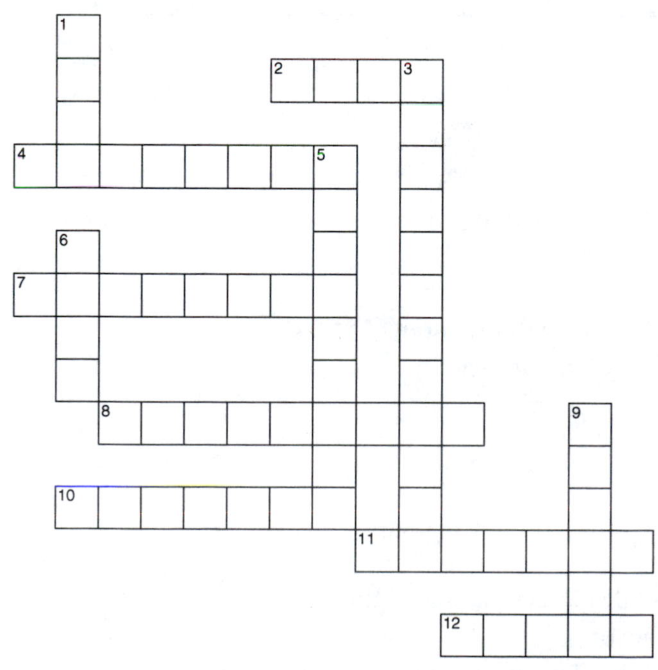

Across

2. to place in sternal recumbancy
4. a two-humped camel
7. a type of alpaca with short, crimped fiber
8. camel with one large hump
10. how alpaca wool density is measured
11. group that includes camels, dromedaries, alpacas and llamas
12. number of chambers in a camelid stomach

Down

1. name for a baby llama or alpaca
3. having a very high body temperature
5. parasites that can cause bone deformities in crias
6. a type of alpaca with long fibers and no crimping
9. cow, or pertaining to cows

UNIT VIII

Chapter 21

Miniature Pigs

1. What are needle teeth and when should they be clipped?

   What: ______________________________

   When: ______________________________

2. What is the best method to use for behavioral training in companion pigs?

   ____

   ____

   With regard to the above, give one example:

   ____

   ____

   ____

   ____

3. What are two of the recommended vaccinations for miniature pigs?

4. Describe Porcine Stress Syndrome.

   ____

   ____

   ____

   ____

Crossword Puzzle

MINIATURE PIGS

Across

2. the canines of newborn piglets
5. a neutered male pig
6. fruit that helps reduce pig odor
8. digging things up with the snout
10. the study of animal behavior

Down

1. most common health problem in pet pigs
3. bacterial disease that causes diamond-shaped skin lesions
4. mature female pig that has had a litter
5. mature male pig
7. check this law before obtaining an agricultural animal
9. females that have not produced a litter